resources

No Nonsense Maths Problems

Book 2

Year 4

Activities to help students apply mathematical knowledge and strategies

Suzi de Gouveia,
Jackie Andrews
and Jude Callaghan

essential
resources

Title:	No Nonsense Maths Problems Activities to help students apply mathematical knowledge and strategies Book 2 – Year 4
Authors:	Suzi de Gouveia, Jackie Andrews and Jude Callaghan
Editor:	Tanya Tremewan
Designer:	Doreen Neason
Book code:	4671
ISBN:	978-1-927190-43-2
Published:	2012
Publisher:	Essential Resources Educational Publishers Limited

United Kingdom:	*Australia:*	*New Zealand:*
Unit 8–10 Parkside Shortgate Lane Laughton, BN8 6DG ph: 0845 3636 147 fax: 0845 3636 148	PO Box 906 Strawberry Hills, NSW 2012 ph: 1800 005 068 fax: 1800 981 213	PO Box 5036 Invercargill ph: 0800 087 376 fax: 0800 937 825

Websites: www.essentialresources.com.au
www.essentialresources.co.nz

About the authors: Suzi is the enthusiastic principal of St Teresa's Primary School in Christchurch who has international teaching experience. She has had the pleasure of teaching in a multi-cultural environment. Thirty years of teaching have enabled Suzi to develop a wealth of ideas and resources to best help children.

Jackie is an experienced teacher who has taught middle primary children in both New Zealand and the United Kingdom. As a mother of four young children she is beginning to wonder where the time to diversify has gone.

Jude is an experienced, enthusiastic teacher with a passion for teaching and learning. Her teaching programmes are innovative and exciting. Jude has embraced the Numeracy Development Programme wholeheartedly. She has joined the writing team to share her deep understanding and wealth of ideas.

Contents

Introduction

No Nonsense Maths Problems has been designed to help busy teachers provide students with activities that reinforce number knowledge and skills taught in their maths programmes. This book is part of a series of problem solving books covering Years 2–4 that supports the teaching of the *Australian Mathematics Curriculum*.

Students need opportunities to decipher information to answer questions and know whether to add or subtract, multiply or divide and, as problems get more complex, use several mathematical operations.

These Year 4 problem solving cards are organised into four sections covering addition and subtraction, multiplication and division, using more than one operation, and fractions. The cards are designed to be used independently, in a group or as a class. Answer cards have been provided at the end of the book. These can be copied and kept with the problem solving cards so students can check each other's work.

To prepare the material in this resource, photocopy the problem solving cards and the answer cards onto coloured paper (we suggest one colour per stage), laminate them and store them in zip-lock bags according to the operation or strand. The cards are then long lasting and ready to be placed into a group box when needed.

No Nonsense Maths Problems complements the *No Nonsense Number Games* and *Maths Works* series to provide a range of enticing worksheets and games that reinforce learning at each level.

Card 1

Addition and subtraction

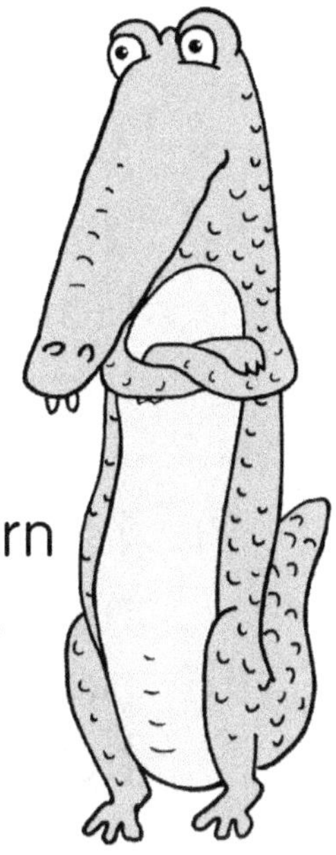

1) Kevin delivers newspapers to homes. He has 1 603 newspapers to deliver but 17 newspapers have been damaged in the rain. How many undamaged newspapers does he have left to deliver?

2) Minki has to sell 75 bars of chocolate to earn enough money to pay for his school camp. He has sold 38 so far. How many more chocolate bars does he need to sell?

No Nonsense Maths Problems Book 2 Year 4

Card 2

Addition and subtraction

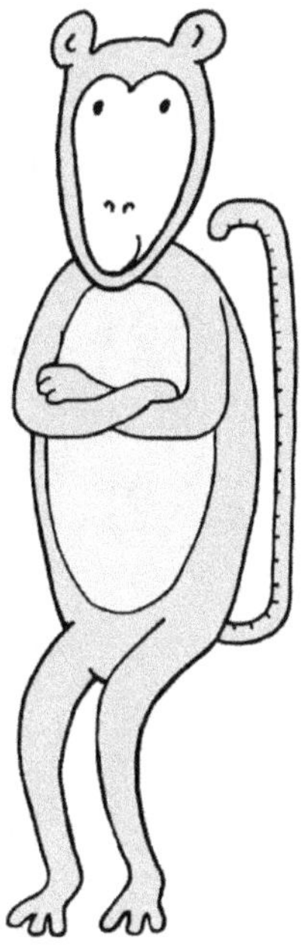

1) Tim has been saving money from his part-time job for the last year. He has $3 206 in his bank account. He withdraws $213 to buy Christmas presents for his family. How much money does he have left in his bank account?

2) Jane buys a bag of balloons for her birthday party. There are 29 blue balloons, 18 green balloons and 37 red balloons in the bag. How many balloons are there altogether?

No Nonsense Maths Problems Book 2 Year 4

Card 3

Addition and subtraction

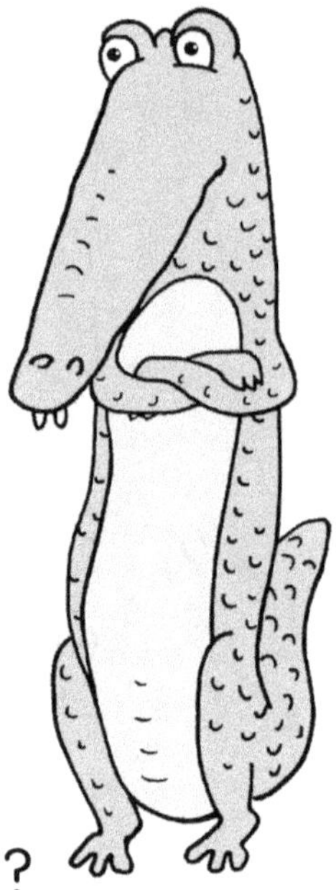

1) 92 people have flown from Sydney to Melbourne. The first 23 passengers have disembarked and collected their bags. How many passengers still need to collect their bags?

2) To get a "Great Achievement" award, Rob needs to swim 94 lengths of the swimming pool. If he has already swum 69 lengths, how many more lengths of the swimming pool does he need to swim to get his award?

No Nonsense Maths Problems Book 2 Year 4

Card 4

Addition and subtraction

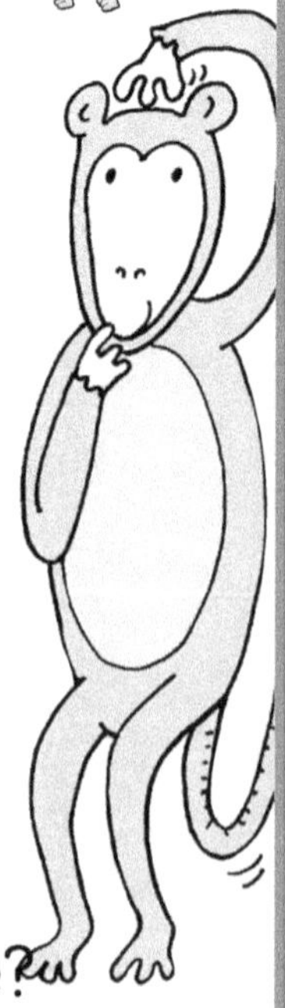

1) Charlotte wants to print out 85 new photos from her computer. If she has already printed 26 of them, how many more photos does she need to print?

2) Kristine had $82 in her purse. After she has bought lunch for her friends she has $54 left in her purse. How much did lunch cost?

3) Leo scores 49 runs in both his first and second innings of a cricket match. What is his total score?

No Nonsense Maths Problems Book 2 Year 4

Card 5

Addition and subtraction

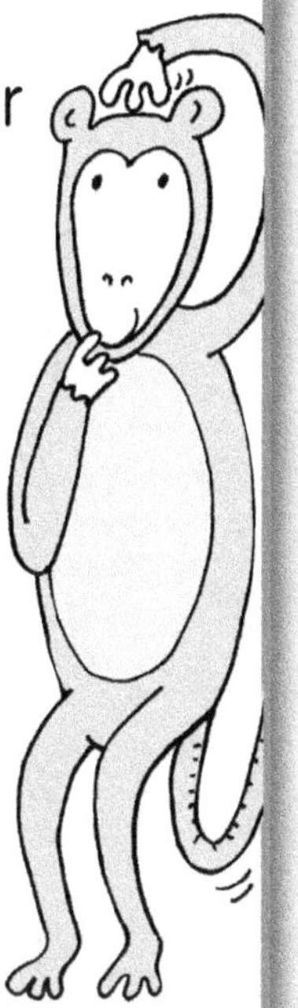

1) Ye Lin's swim club raises $652 during the summer holidays. In winter it has a swim-a-thon which brings the total to $954 across the two seasons. How much money does the club raise from the swim-a-thon?

2) During the first term, Room 9 raises $584. By the end of the second term it has raised $845 in total. How much money does the class raise in the second term?

No Nonsense Maths Problems Book 2 Year 4

Card 6

Addition and subtraction

1) Jack saved $210. He spent some money and has $148 left. How much money did he spend?

2) On Monday Thea reads her book from page 34. On Friday she gets up to page 425. How many pages has she read?

3) Phoenix's holiday job is pruning grapevines. He had pruned 23 of the 92 vines that he needs to do this week. How many does he still have to do?

No Nonsense Maths Problems Book 2 Year 4

Card 7

Addition and subtraction

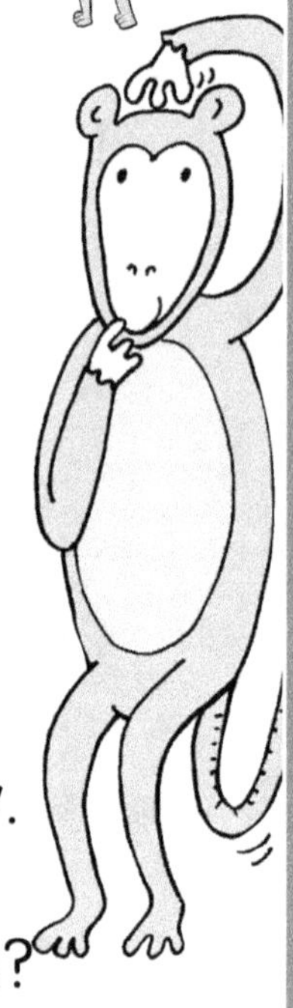

1) Jim has handed out 598 of his 765 flyers. How many flyers does he still need to hand out?

2) John's team has scored 147 runs in its first innings of a cricket match. It scores a total of 589 runs across both innings. How many runs does the team get in the second innings?

3) Sarah bakes 84 cupcakes for her birthday party. After the party there are 19 cupcakes left. How many cupcakes have she and her friends eaten?

No Nonsense Maths Problems Book 2 Year 4

Card 8

Addition and subtraction

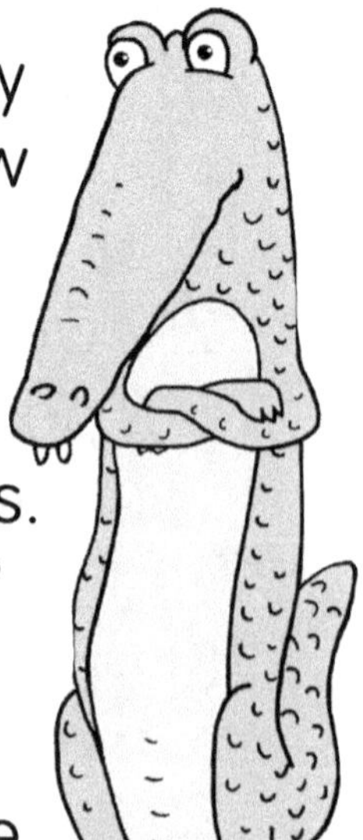

1) Min Joo's father works in a town 99 km away and he drives there and back each day. How many kilometres does he drive in a day?

2) 37 people have bought tickets for a play. The theatre must sell 92 tickets to cover costs. How many more tickets does it need to sell?

3) Cedric catches the bus to school each day. He travels 57 km each way. How far does he travel in a day?

No Nonsense Maths Problems Book 2 Year 4

Card 9

Addition and subtraction

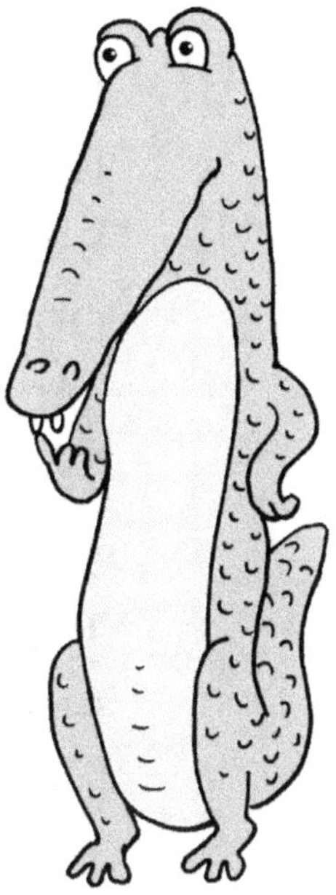

1) Ryan has swum 27 laps of the pool. He has to do 74 in total. How many more laps does he still have to do?

2) Claudia had $463 in her bank account. Then she bought a bike for $298. How much does she have left in her account?

3) Brooke is doing a training bike ride that is 94 km long. She has already biked 67 km. How far does she still have to go?

Card 10

Addition and subtraction

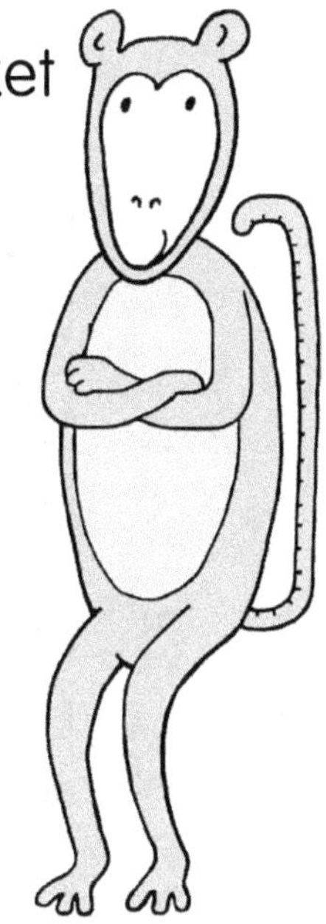

1) Elliot saves $30 and gets another $23 in pocket money. He spends $7 on a calculator, $7 on a pencil case and $38 on a pen drive. How much money does Elliot have left?

2) Caitlin has a holiday job working for a local deli. Her job is to peel apples for pies. She has peeled 26 of the 85 apples she needs to do today. How many more apples does she still need to peel?

No Nonsense Maths Problems Book 2 Year 4

Card 11

Addition and subtraction

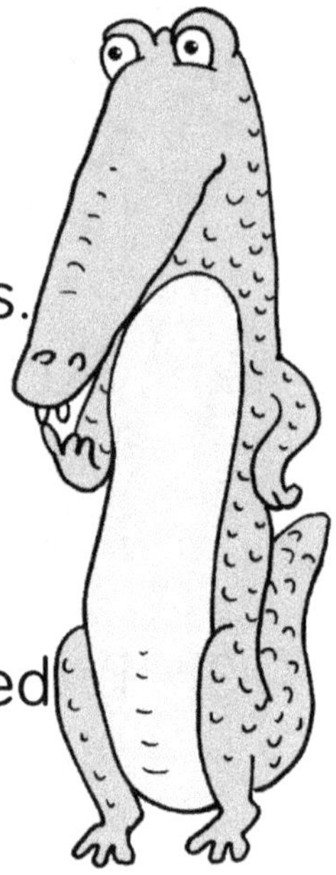

1) Hayden's maths teacher gives his class a maths test four days a week and gives students 1 mark for each question they answer correctly. Here are Hayden's scores.

Monday	23
Tuesday	44
Wednesday	47
Thursday	36

How many questions has Hayden answered correctly this week?

No Nonsense Maths Problems Book 2 Year 4

Card 12

Addition and subtraction

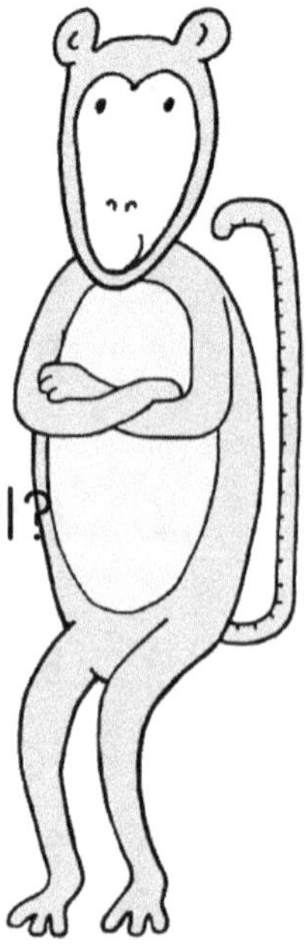

1) Naomi picks raspberries for four days.

Day 1	19 raspberries
Day 2	54 raspberries
Day 3	31 raspberries
Day 4	36 raspberries

How many raspberries has she picked in total?

2) Noah is reading a 624-page book. He has read 379 pages. How many more pages does he need to read?

No Nonsense Maths Problems Book 2 Year 4

Card 13 Addition and subtraction

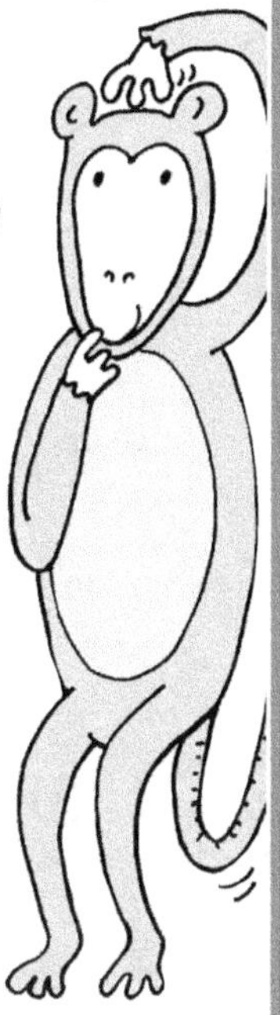

1) Ethan is saving up to buy a new surfboard that costs $978. He has saved $195. How much more does he need to save?

2) Fred had $536 in his bank account. Then he bought a skateboard for $399. How much does he have left in his account?

3) Sam is saving up to buy a new tennis racquet that costs $633. She has saved $258 so far. How much more does she need to save?

No Nonsense Maths Problems Book 2 Year 4

Card 14 Addition and subtraction

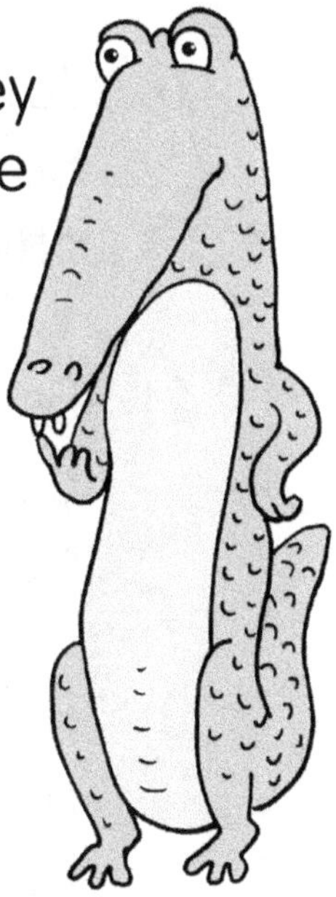

1) Riley wants to buy a new pair of runners. They cost $95. He has saved $32. How much more does he need to save?

2) Francis has been recording how many runs he has scored in cricket over the last month.

Week 1	45 runs
Week 2	18 runs
Week 3	42 runs
Week 4	25 runs

How many runs has he made in total?

No Nonsense Maths Problems Book 2 Year 4

Card 15

Addition and subtraction

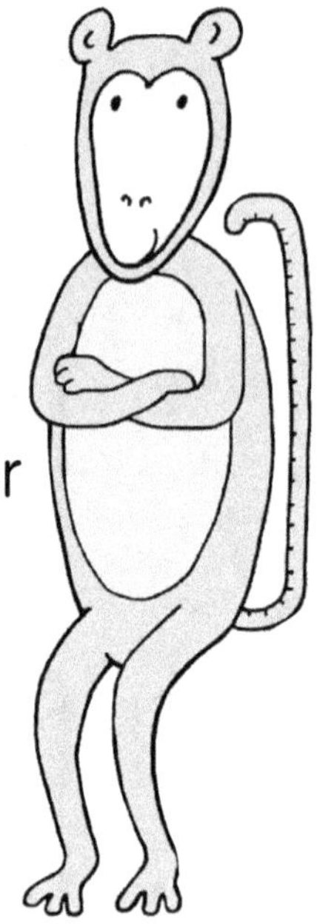

1) Lexi had $654 in her bank account. Then she bought an iPod for $298. How much money does she have left in her account?

2) Xavier drives 563 km in one week and Miah drives 376 km. How much further does Xavier drive than Miah?

3) Gracie is saving to buy a new MP3 player that costs $359. She has saved $281. How much more does she need to save?

No Nonsense Maths Problems Book 2 Year 4

Card 16

Addition and subtraction

1) Jane wants to buy a new pair of jeans costing $87. She has saved $16. How much more does she need to save?

2) Levi had $654 in his bank account. Then he bought a basketball for $98. How much does he have left in his bank account?

3) Rob had $463 in his bank account. Then he bought a phone for $197. How much does he have left in his bank account?

No Nonsense Maths Problems Book 2 Year 4

Card 17

Addition and subtraction

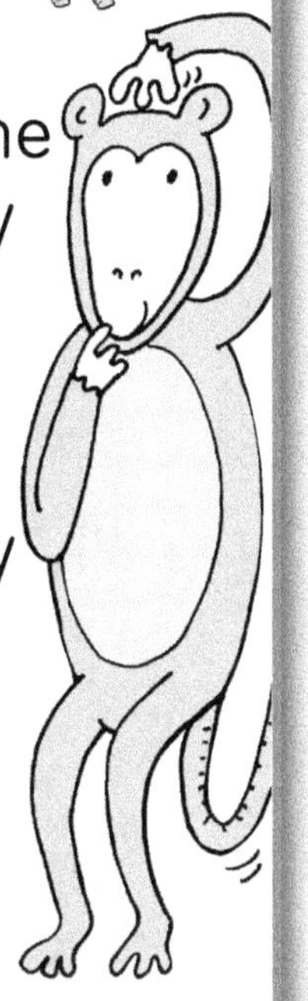

1) Charlotte had $505 in her bank account. Then she bought a new phone for $197. How much money does she have left in her account?

2) Tim had $505 in his bank account. Then he bought a pair of skis for $399. How much money does he have left in his bank account?

3) Greg is saving for a new surfboard that costs $515. How much more money does he need to save if he has already saved $346?

No Nonsense Maths Problems Book 2 Year 4

Card 18

Addition and subtraction

1) Sarah and her parents are going on holiday. They travel 176 km on Monday, 250 km on Tuesday and 195 km on Wednesday. How far have they travelled altogether?

2) Caitlin's mum bought her a new basketball and Caitlin needs to pay her back. The ball cost $53 and Caitlin has saved $18 so far. How much more does she need to save?

No Nonsense Maths Problems Book 2 Year 4

Card 1

Multiplication and division

1) There are 36 strawberries in a bowl to be shared equally among a group of people. If each person eats 9 strawberries, how many people are there altogether?

2) Martha walks her dog for 25 minutes each day. How many minutes does she spend walking her dog in a week?

3) How many boxes can be wrapped in 48 minutes if it takes 6 minutes to wrap 1 box?

No Nonsense Maths Problems Book 2 Year 4

Card 2

Multiplication and division

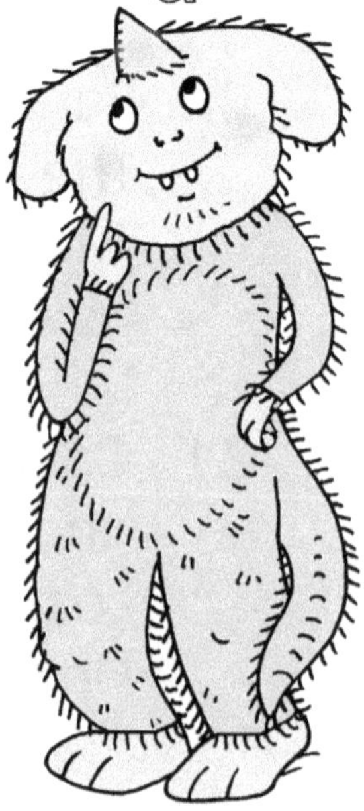

1) Luke buys 7 car magazines for $56, with each magazine costing the same amount. If Leo buys 3 of them from Luke for the same price per magazine, how much does he pay Luke?

2) There are 48 pieces of fruit to be shared equally among a group of people. If each person eats 6 pieces, how many people are there altogether?

No Nonsense Maths Problems Book 2 Year 4

Card 3

Multiplication and division

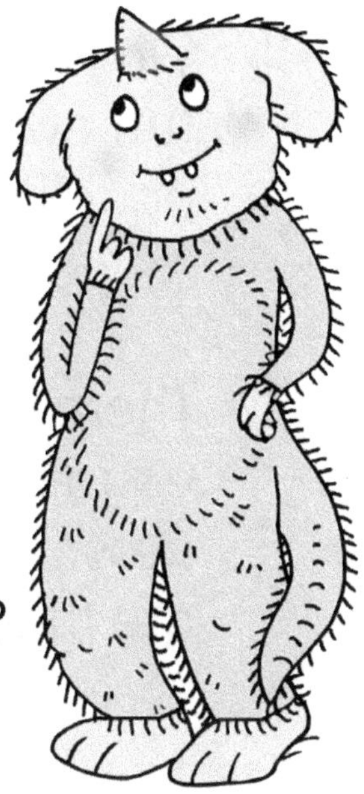

1) A netball team enters a tournament. The tournament costs $16 per person. If there are 9 girls in the team, how much will the team pay in total?

2) Ye Lin has 5 bookshelves in her bedroom, containing a total of 45 books. Each bookshelf contains the same number of books. How many books are on each shelf?

No Nonsense Maths Problems Book 2 Year 4

Card 4

Multiplication and division

1) It takes Thea 5 minutes to wash the breakfast dishes each day. How long does she spend doing the breakfast dishes in total each week?

2) There are 48 pieces of sushi to share equally among 6 people. How many pieces of sushi will each person get?

3) How many hard-boiled eggs can be peeled in 18 minutes if it takes 2 minutes to peel 1 egg?

No Nonsense Maths Problems Book 2 Year 4

Card 5

Multiplication and division

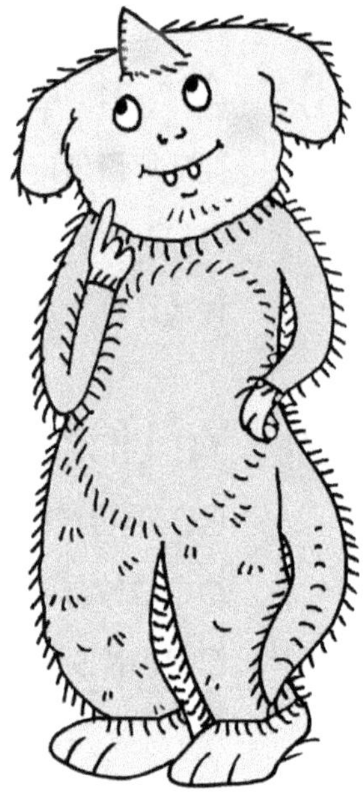

1) Waverly School raises $4 200 at its family fun evening. The money is shared out equally among 6 classes. How much money does each class get?

2) There are 49 olives in a jar to be shared equally among a group of people. If each person eats 7 olives, how many people are there altogether?

No Nonsense Maths Problems Book 2 Year 4

Card 6

Multiplication and division

1) Jack has earned $256 for working in his dad's shop. He is paid in ten dollar notes and one dollar coins. What is the maximum number of notes he could get?

2) Noah slept for a total of 63 hours last week. If he slept the same number of hours each night, how many hours did he spend asleep each night?

No Nonsense Maths Problems Book 2 Year 4

Card 7

Multiplication and division

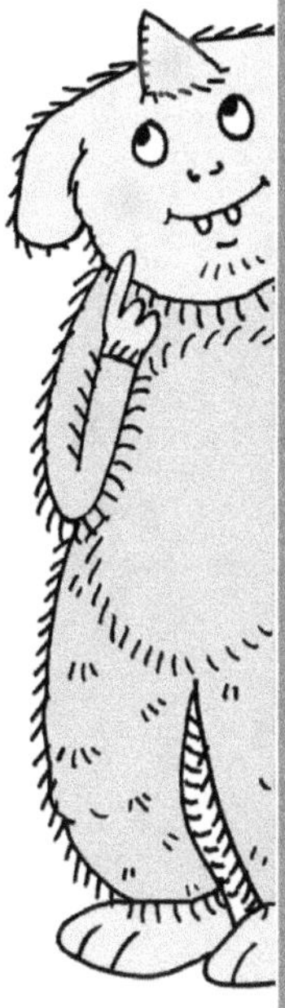

1) The teacher organises his 52 students into 4 equally sized groups. How many students are in each group?

2) Jim and 4 friends each plant 62 native trees on Conservation Day. How many trees do they plant in total?

3) Min Joo bakes 4 batches of 32 muffins. How many muffins does she bake in total?

Card 8

Multiplication and division

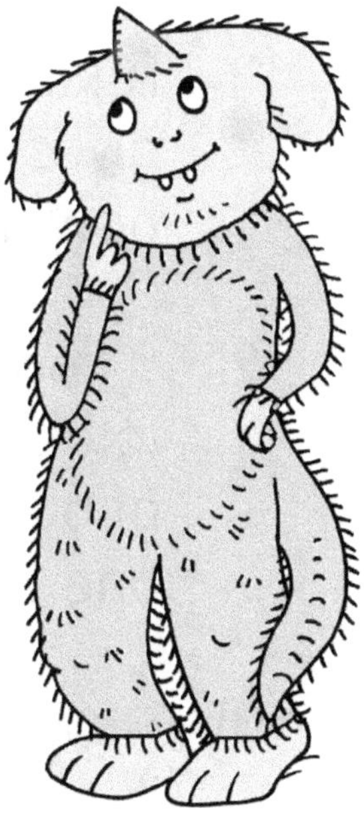

1) Claudia makes 69 truffles and divides them equally among 4 bags. How many does she put into each bag and how many are left over?

2) The 7 boys in John's all-boy band each pay $79 towards the travel costs on their music tour. How much does the band pay in total?

Card 9

Multiplication and division

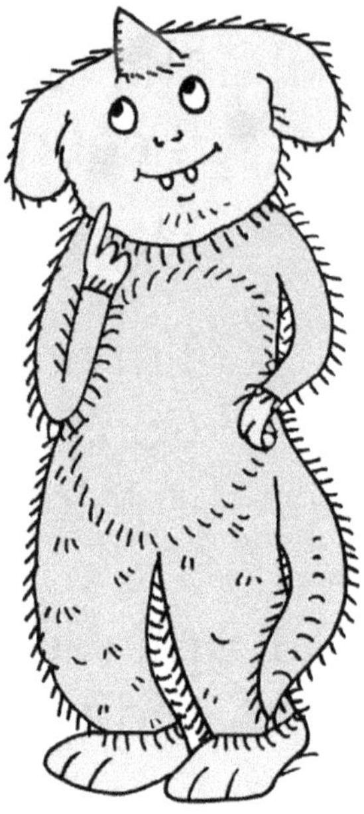

1) Tickets for the school dance cost $10 each. So far 135 people have bought tickets. How much money has been paid in total?

2) Minki makes 88 flyers and gives an equal number to each of his 4 friends to hand out. How many flyers does each friend need to hand out?

3) Cedric buys 6 books worth $42 each. How much does he spend in total?

No Nonsense Maths Problems Book 2 Year 4

Card 10

Multiplication and division

1) Luke has 174 boxes to deliver to 6 shops. If he gives each shop the same number of boxes, how many will each shop get?

2) 9 florists arrange 108 bunches of flowers in total. Each one arranges the same number of bunches. How many bunches does each one arrange?

3) How many apples can you eat in 28 minutes if it takes you 4 minutes to eat 1 apple?

No Nonsense Maths Problems Book 2 Year 4

Card 11

Multiplication and division

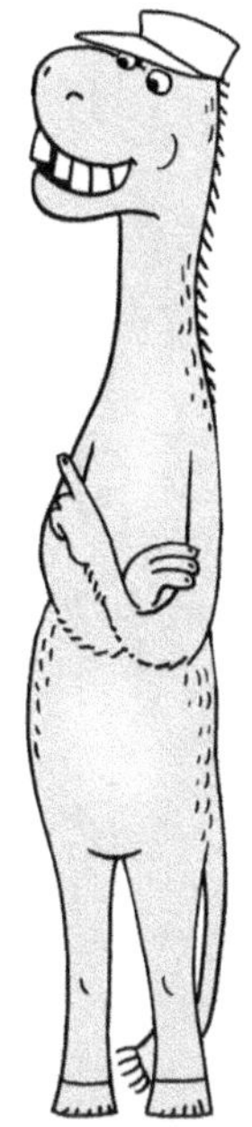

Elliot, Caitlin and their mum and dad are going on a plane trip.

1) A flight costs $198 one way for each person. How much will the family need to pay to get to their destination and back again?

2) The accommodation costs $799 per person. What will it cost in total?

Card 12

Multiplication and division

1) Luke has to buy some stationery for his dad – 22 reams of paper that cost $7 each. How much money does his father owe him?

2) Phoenix and his three friends spend Friday night busking in town. They make $396 between them and share out this total equally. How much does each of them get?

No Nonsense Maths Problems Book 2 Year 4

Card 13

Multiplication and division

1) Jade has made 133 greetings cards and has to store the same number of cards in each of the 7 boxes. How many does she put into each box?

2) Naomi walked 42 km last week. If she walked the same distance each day, how many kilometres did she walk each day? If Naomi walks the same distance for 9 weeks, how far will she walk in total?

No Nonsense Maths Problems Book 2 Year 4

Card 14

Multiplication and division

1) Bella swam 28 km last week. If she swam the same distance each day, how many kilometres did she swim in one day? If Bella swims the same distance for 9 weeks, how far will she swim in total?

2) 7 teams enter a soccer competition. There are 11 members in each team. Each team member has to pay $10 towards the entrance fee. How much did all the teams pay in total?

No Nonsense Maths Problems Book 2 Year 4

Card
15
Multiplication and division

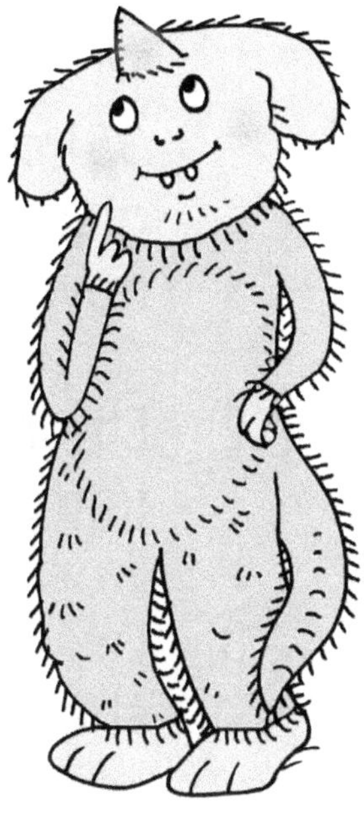

1) There are 203 birds in the Downtown Aviary. The head keeper puts the same number of birds in each of the 7 cages. How many birds are in each cage?

2) Francis ran for 49 km last week. If he ran the same distance each day, how many kilometres did he run in one day? If Francis runs the same distance for 9 weeks, how far will he run in total?

Card
16
Multiplication and division

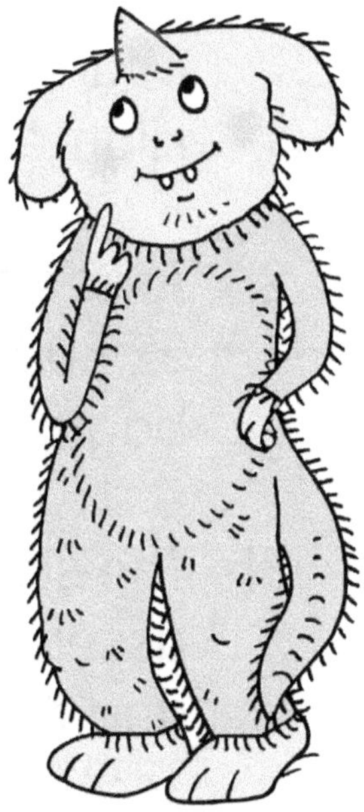

1) The teacher organises her 64 students into 4 equally sized groups. How many students are in each group?

2) The 9 members of Belinda's netball team have raised $3 600 to help them get to a tournament. They share it out equally so that each of them has money to pay towards their own plane ticket. How much does each team member get?

No Nonsense Maths Problems Book 2 Year 4

Card 17 Multiplication and division

1) The 9 members of Gracie's netball team make 351 cheese rolls as a fundraiser. If each of the girls makes the same number of cheese rolls, how many does each girl make?

2) Ethan's rugby club buys 9 kit bags for $2 700 in total. How much does each kit bag cost?

3) Lexi saves $41 each week. How much does she save in 4 weeks?

No Nonsense Maths Problems Book 2 Year 4

Card 18 Multiplication and division

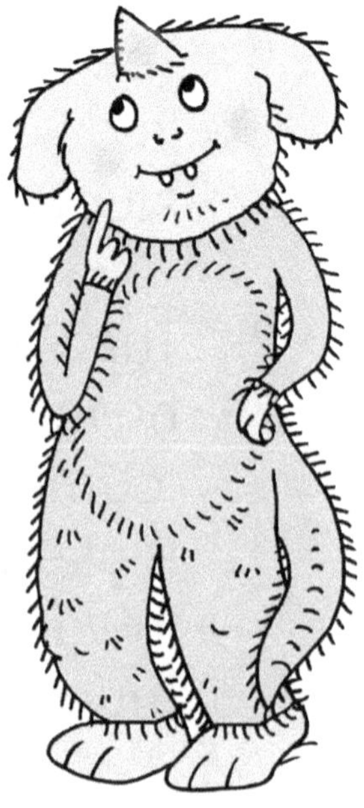

1) Miah walks 9 km each day. How many kilometres will she walk in one week?

2) Jane's soccer club buys 7 kit bags for $2 800 in total. How much does each kit bag cost?

3) The school hall has been set up for assembly. There are 36 seats in each row and there are 8 rows in total. How many seats are there altogether?

No Nonsense Maths Problems Book 2 Year 4

Card 19

Multiplication and division

1) Riley saves $14 each week. How much does he save in 8 weeks?

2) There are 52 windows to go into a building. Each storey must have 4 windows exactly. How many storeys does the building have in total?

3) There are 8 classes in Year 6. Each class contains 31 students. How many Year 6 students are there in total?

Card 20

Multiplication and division

1) The 4 members of Sarah's drama club have to pay $12 each for costumes. How much do the costumes cost in total?

2) There are 32 students in Levi's class. He has 8 pies to share with them all. How many pieces should he cut each pie into so that everyone can have one piece?

3) Ye Lin saves $16 each week. How much does she save in 6 weeks?

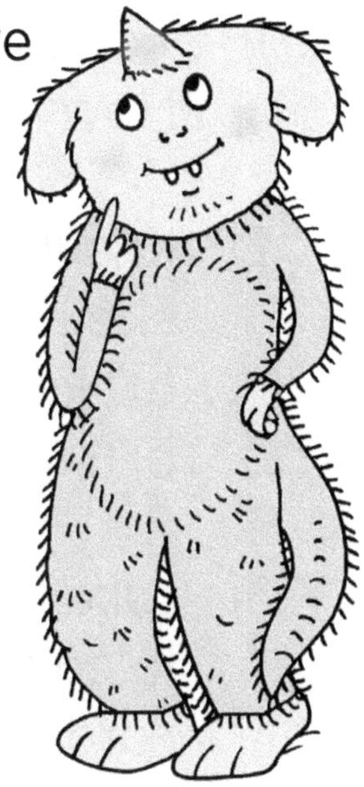

Card 21 Multiplication and division

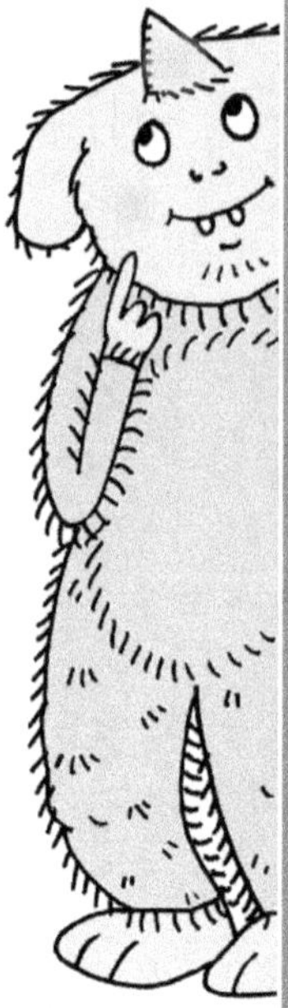

1) The farmer puts 61 sheep in each of the 5 pens. How many sheep does he pen up in total?

2) Charlotte's school has bought interactive whiteboards on a special deal: they get 4 whiteboards for $2 400. How much did each whiteboard cost?

3) In a bike race, Tim and his 4 friends have to ride 22 km each. How far do they travel in total?

Card 22 Multiplication and division

1) Xavier runs 7 km each day of the week. How many kilometres will he run in one week?

2) Kristine is training for a bike race. For 4 weeks she has to bike 36 km each day.
a) How far does she bike in a week?
b) How far does she bike in 4 weeks?

3) Luke has 54 rugby cards. He gives 4 cards to each of his 6 friends. How many rugby cards does he have left?

No Nonsense Maths Problems Book 2 Year 4

Card 23 Multiplication and division

1) Hayden's school has bought laptops on a special deal: they get 5 laptops for $2 500. How much did each laptop cost?

2) Eight families dig shellfish together and gather 104 shellfish in total. They share out the total equally. How many shellfish does each family get?

3) Rob saves $19 in one week. How much does he save in 9 weeks?

Card 24 Multiplication and division

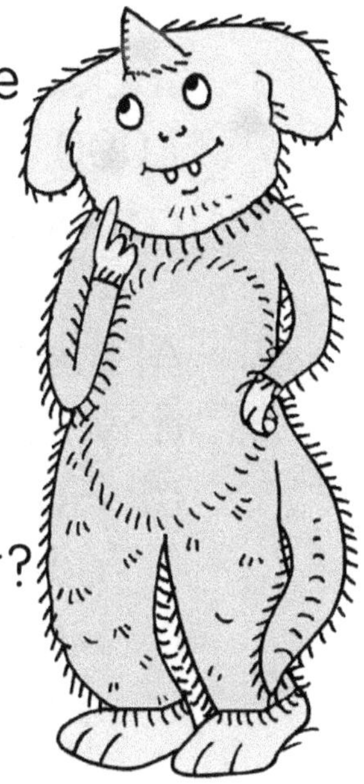

1) Thea grows 35 rows of parsnips. There are 1 000 parsnips in each row. How many parsnips has she grown altogether?

2) Martha makes 96 flyers and gives an equal number to each of her 8 best friends to hand out. How many flyers does each of her friends need to hand out?

3) Robert saves $31 each week. How much does he save in 7 weeks?

No Nonsense Maths Problems Book 2 Year 4

Card 25 Multiplication and division

1) The school is raising money to buy a class set of calculators. It wants to buy 32 calculators at $8 each. How much money does it need to raise?

2) Leo grows 32 rows of cabbages. There are 100 cabbages in each row. How many cabbages has he grown altogether?

3) Jim makes 67 toffees and divides them equally into 8 bags. How many toffees does he put into each bag and how many are left over?

No Nonsense Maths Problems Book 2 Year 4

Card 26 Multiplication and division

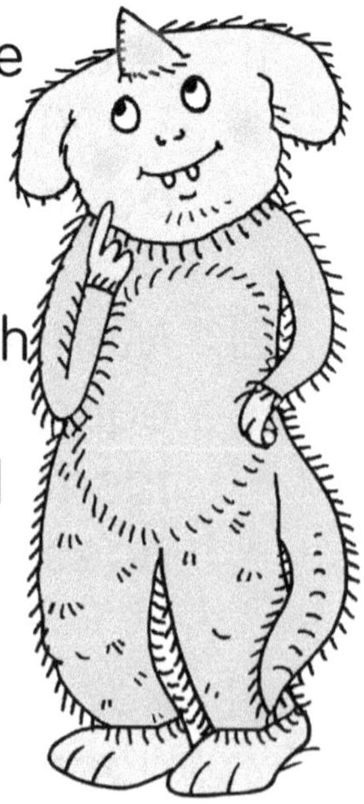

1) Jack grows 96 rows of cabbages. There are 1 000 cabbages in each row. How many cabbages has he grown altogether?

2) At the zoo each lion eats 16 kg of meat each day. The zoo has 6 lions.
   a) How much meat do the lions eat in total each day?
   b) How much meat do they eat in total each week?

No Nonsense Maths Problems Book 2 Year 4

Card 27

Multiplication and division

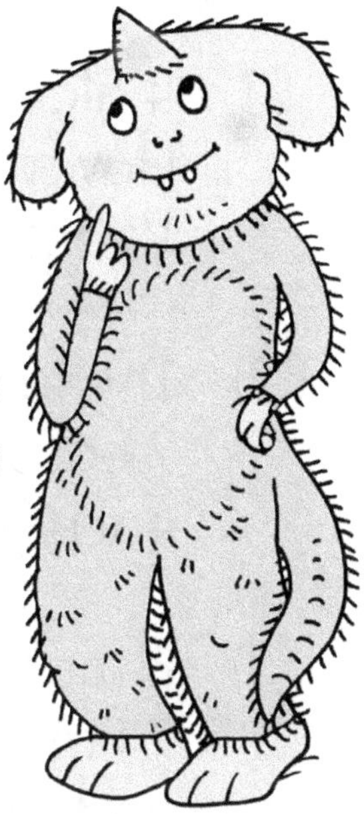

1) Claudia grows 84 rows of carrots. There are 100 carrots in each row. How many carrots has she grown altogether?

2) The teacher organises her 32 students into 8 equal groups. How many students are in each group?

3) Luke makes 96 toffees and divides them equally among 4 bags. How many toffees does he put into each bag?

No Nonsense Maths Problems Book 2 Year 4

Card 28

Multiplication and division

1) John plants 179 000 radish seedlings. There are 100 radishes in each row. How many rows of seedlings are there?

2) Each of the 7 girls in Brooke's netball team raises $92. How much does the team raise in total?

3) Min Joo picks 49 plums. She puts 6 plums into each of her 7 jars. How many plums does she have left over?

No Nonsense Maths Problems Book 2 Year 4

Card 29

Multiplication and division

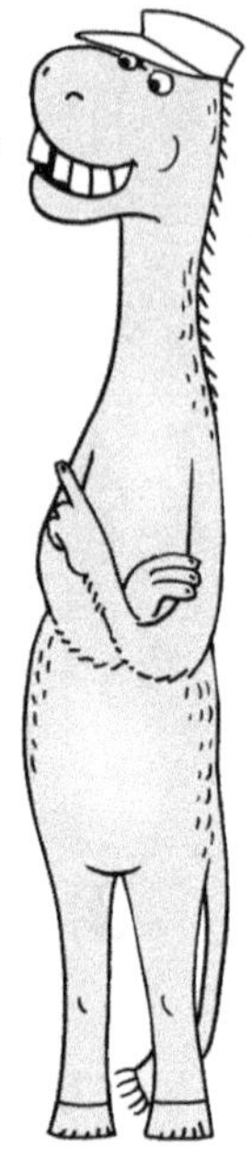

1) 4 glasses of milk can be poured from each bottle. How many glasses of milk can be poured from 18 bottles?

2) The teacher organises his 42 students into 7 equal groups. How many students are in each group?

3) 6 mangos are packed in each bag. If there are 74 mangos, how many full bags can be packed and how many mangos are left over?

No Nonsense Maths Problems Book 2 Year 4

Card 30

Multiplication and division

1) The baker bakes 6 trays of chocolate chip cookies with 18 cookies on each tray. How many cookies does he bake in total?

2) Caitlin plants 65 000 strawberry plants. There are 65 rows of strawberry plants. How many strawberry plants are in each row?

3) It takes Jade 12 minutes to wash 3 windows. How long will it take her to wash 14 windows?

No Nonsense Maths Problems Book 2 Year 4

Card 31

Multiplication and division

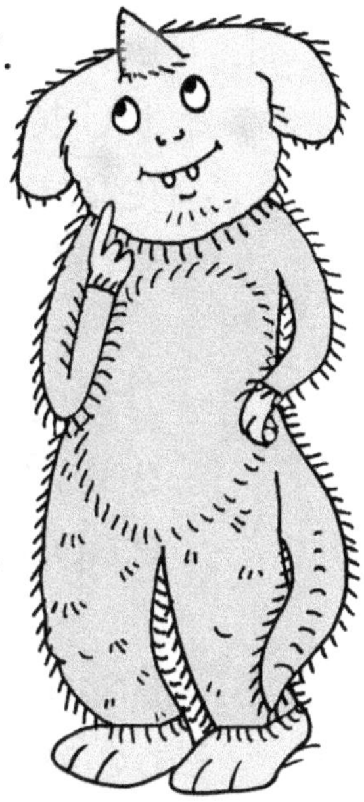

1) There are 8 mini chocolate bars in a packet. If there are 17 packets left in the shop, how many mini chocolate bars is this?

2) 8 buses have 24 seats each. How many seats is this altogether?

3) Minki plants 320 000 cauliflower seedlings. There are 32 rows of cauliflower seedlings. How many cauliflower seedlings are in each row?

No Nonsense Maths Problems Book 2 Year 4

Card 32

Multiplication and division

1) There are 37 tickets to share evenly among 5 friends. How many tickets does each friend get and how many are left over?

2) Cedric makes 48 flyers and gives an equal number to each of his 8 friends to hand out. How many flyers does each friend need to hand out?

3) A sports bag holds 7 balls. How many sports bags are needed for 49 balls?

No Nonsense Maths Problems Book 2 Year 4

Card 33 Multiplication and division

1) Each student doing science badges brings $10 to contribute to the cost of a field trip. Their teacher collects $570 altogether. How many students will be going on the trip?

2) There are 9 elephants in the zoo. If each elephant eats 16 kg of hay a day, how much do the elephants eat in one day altogether?

3) Brooke delivers 82 papers each day. How many papers does she deliver in 5 days?

No Nonsense Maths Problems Book 2 Year 4

Card 34 Multiplication and division

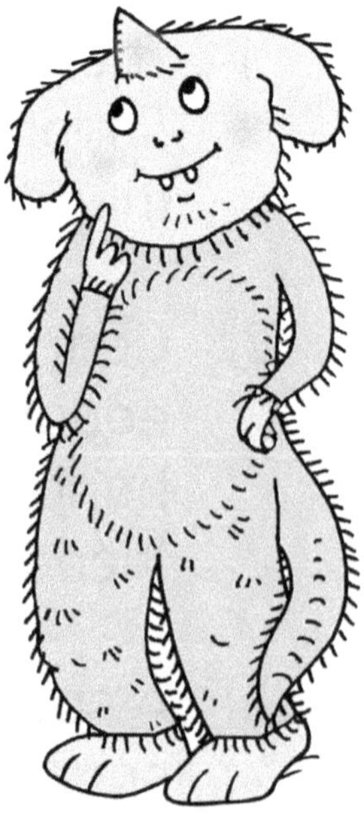

1) Bella makes 68 flyers and gives an equal number to each of her 4 best friends to hand out. How many does each friend need to hand out?

2) Kelly earns $18 a month delivering her local newspaper. How much money does she earn in 7 months?

3) If there are 9 biscuits in each bag, how many bags are needed for 540 biscuits?

No Nonsense Maths Problems Book 2 Year 4

Card 1

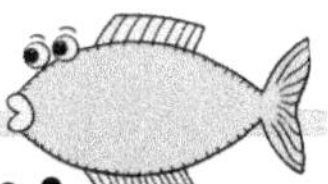

Using more than one operation

1) Elliot has 5 more tennis balls than Jo. Phoenix has 7 times as many tennis balls as Elliot. If Phoenix has 56 tennis balls, how many does Jo have?

2) Tim, Jane, Tom and Sue together earn $84 by cleaning cars. They need $12 of this total for supplies and then share the remaining money equally. How much does each person get?

No Nonsense Maths Problems Book 2 Year 4

Card 2

Using more than one operation

1) There are 28 bicycles and 24 cars in the garage at Jo's apartment building. How many wheels are in the garage?

2) Gracie has 5 boxes, each containing 24 chocolates. How many chocolates does she have altogether? She takes them on school camp to give away equally among her 20 friends. How many chocolates will each friend get?

No Nonsense Maths Problems Book 2 Year 4

Card 3 Using more than one operation

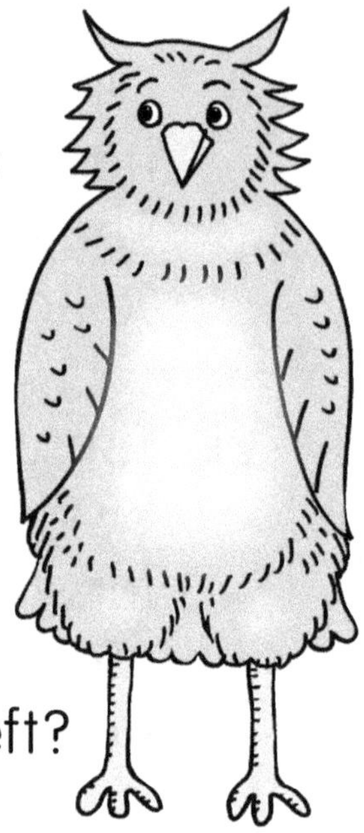

1) Lexi makes 77 bags of sherbet to sell at the market. She sells 59 bags. How many bags does she still have to sell? If Lexi sells each bag for $2, how much money does she make altogether from 59 bags?

2) Thea has $21 in savings and she gets another $14 for pocket money. She then spends $3 on a magazine and $31 on a book. How much money does Thea have left?

No Nonsense Maths Problems Book 2 Year 4

Card 4 Using more than one operation

1) The bus that Tom and Sue's team hires to get to the touch rugby tournament costs $171. The families of the 9 team members share the cost equally. Tom and Sue are brother and sister so their family has to pay for two. How much does their family pay?

2) There are 8 bicycles and 6 tricycles in the storage area at Hayden's apartment building. How many wheels are there in all?

No Nonsense Maths Problems Book 2 Year 4

Card 5

Using more than one operation

1) A bag of apples costs $3 and a bunch of bananas costs $2. The school tuck shop buys 7 bags of apples and 4 bunches of bananas. How much does the tuck shop spend on fruit altogether?

2) Sarah buys 3 packs of red clips, 8 packs of yellow clips and 3 packs of green clips. There are 19 clips in each pack. How many clips does Sarah buy in all?

No Nonsense Maths Problems Book 2 Year 4

Card 6

Using more than one operation

1) Charlotte's aunt gives her $172 to spend on sports gear. Charlotte buys 4 singlets that cost $13 each, a netball that costs $25 and a pair of shorts that costs $18. How much money does Charlotte have left after she has bought all this gear?

2) Todd needs 64 paper plates for a birthday party. He already has 6 blue plates and 31 red plates. How many more plates should Todd buy?

No Nonsense Maths Problems Book 2 Year 4

Card 7

Using more than one operation

1) Ye Lin buys a sandwich for $5.70, a bag of chips for $2.50 and a drink for $0.80. She gives the teller $10.00. How much change should Ye Lin receive?

2) Miah packs apples into bags for her holiday job. She puts 10 apples into each bag. By the end of the day she has filled 62 bags and has 9 apples left over. How many apples did she have for packing at the beginning of the day?

No Nonsense Maths Problems Book 2 Year 4

Card 8

Using more than one operation

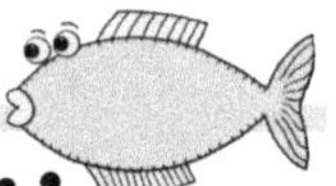

1) Noah buys 3 packages of cat food and 2 packages of dog food. Each package contains 9 tins. How many tins of pet food does Noah buy?

2) Bella needs 48 cupcakes for a birthday party. She has already made 12 chocolate cupcakes and 18 strawberry cupcakes. How many more cupcakes does Bella need to make?

No Nonsense Maths Problems Book 2 Year 4

Card
9

Using more than one operation

1) Claudia is collecting coins. She gets 5 coins from her piggy bank and 3 coins from her brother. Her father gives her 6 coins. Then Claudia gives 5 of her coins to her friend Francis. How many coins does Claudia have left?

2) Brooke buys 12 bags of red apples and 9 bags of yellow apples. Each bag contains 7 apples. How many apples does Brooke buy altogether?

No Nonsense Maths Problems Book 2 Year 4

Card
10

Using more than one operation

1) Caitlin walks 4 blocks from her house to the bus stop. She rides the bus 3 blocks to the library. Later she comes home the same way. How many blocks does Caitlin travel in all?

2) Martha has $25 in her piggy bank and saves another $17 from her pocket money. Then she spends $4 on a comic book and $8 on a puzzle. How much money does Martha have left?

No Nonsense Maths Problems Book 2 Year 4

Card 11

Using more than one operation

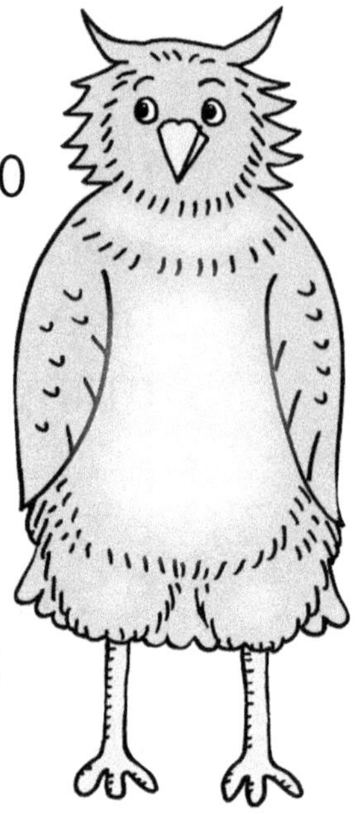

1) Ethan goes to a pizza restaurant with 2 friends. They order a large pizza for $27.70 and a salad for $7.30. They also get 2 cans of drink for $1.40 each. How much money do they spend in total?

2) Jade had 11 marbles. Then she bought another 7 bags of marbles. There are 4 marbles in each bag. How many marbles does Jade have now?

No Nonsense Maths Problems Book 2 Year 4

Card 12

Using more than one operation

1) Levi's uncle gives him $142 to spend on rugby gear. Levi buys 3 jerseys that cost $13 each, a rugby ball that costs $25 and a pair of shorts that costs $16. How much money does Levi have left?

2) Riley buys 9 books about plants and 4 books about fish. He also buys 4 rugby magazines. Each book costs $6 and each magazine also costs $6. How much does Riley spend in all?

No Nonsense Maths Problems Book 2 Year 4

Card 13

Using more than one operation

1) Xavier's aunt gives him $157 to spend on clothes at the mall. He buys 2 t-shirts that cost $15 each and a pair of jeans that costs $28. How much money does Xavier have left to buy more clothes?

2) Jo buys 7 tubs of ice cream and 3 pottles of yoghurt. Each tub of ice cream costs $8 and each pottle of yoghurt costs $2. How much more does Jo spend on ice cream than on yoghurt?

Card 14

Using more than one operation

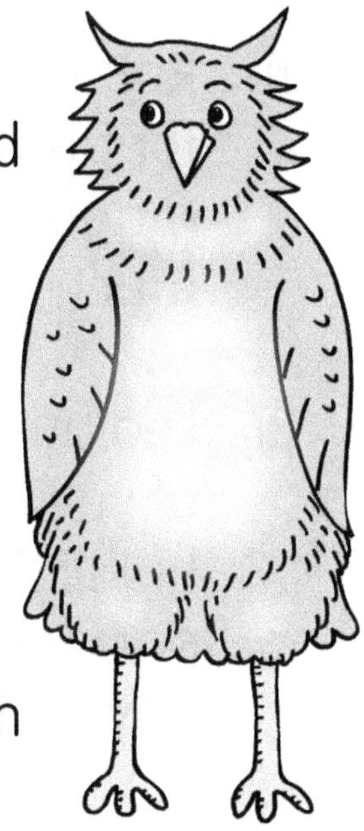

1) For a school fundraiser, Naomi needs to sell 25 chocolate bars. So far, she has sold 3 bars to her grandmother, 3 bars to her uncle and 9 bars to a neighbour. How many more chocolate bars does Naomi need to sell?

2) 16 students and 3 teachers go on a trip to the museum. Student tickets cost $4 each and adult tickets cost $10 each. How much do the museum tickets cost in total?

Card 15 Using more than one operation

1) Tom saved $55 and he got another $13 for pocket money. Then he spent $10 on a book, $12 on a DVD and $31 on a t-shirt. How much money does Tom have left?

2) Kristine buys 7 books about sport and 6 books about dinosaurs. She also buys 8 magazines. Each book costs $4 and each magazine costs $6. How much does Kristine spend in all?

Card 16 Using more than one operation

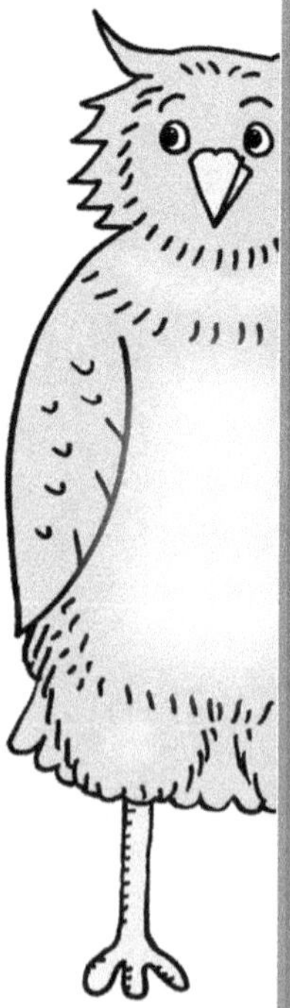

1) Rob had $11 in savings and he got another $18 for pocket money. Then he spent $6 on a magazine and $19 on a computer game. How much money does Rob have left?

2) Leo buys 4 packs of red highlighters, 9 packs of yellow highlighters and 6 packs of green highlighters. There are 6 highlighters in each pack. How many highlighters does Leo buy in all?

Card 1

Fractions

1) There are 56 vehicles in the car park. Six eighths of the vehicles are cars and the rest are trucks.
   a) How many cars are there?
   b) How many trucks are there?

2) Tim and Jane both eat some popcorn from a box they have bought to share. Tim eats one quarter and Jane eats one eighth of the popcorn. Who has eaten more?

Card 2

Fractions

1) Ye Lin is helping clean up her local beach. She and her neighbours collect 24 bags of rubbish in total. 8 of the bags are filled with only plastic. What fraction of the total number of filled bags is this?

2) There are 360 cars in a car park. Four ninths of the cars are red and one third of the cars are white.
   a) How many cars are white?
   b) How many cars are red?

No Nonsense Maths Problems Book 2 Year 4

Card 3 Fractions

1) Tina completes three quarters of an 800 m run. How far does she run?

2) Hayden makes 27 model planes. He paints 3 red. What fraction of the planes is red?

3) There are 48 ice creams in the freezer. Luke and his friends eat two eighths of them. Gracie and her friends eat five eighths. How many ice creams are left over?

No Nonsense Maths Problems Book 2 Year 4

Card 4 Fractions

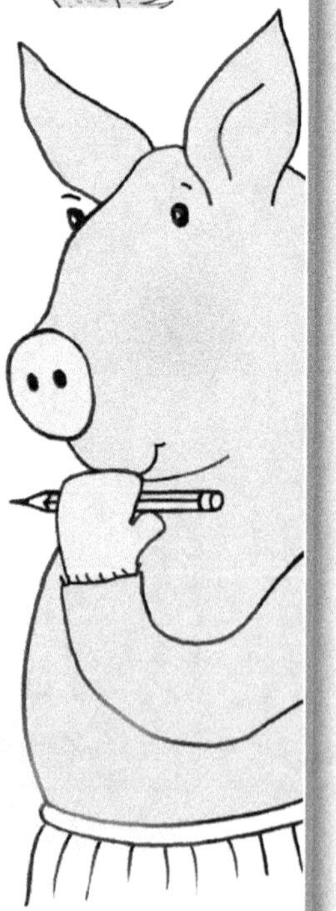

1) Miah makes 18 muffins. She freezes one third of them and puts the rest into a tin. How many muffins does she freeze?

2) There are 36 students in Elliot's art class. One quarter of them have completed their latest project. How many students is this?

3) Phoenix saves $210. If he spends three sevenths of his savings on a computer game, how much does he spend?

No Nonsense Maths Problems Book 2 Year 4

Card 5

Fractions

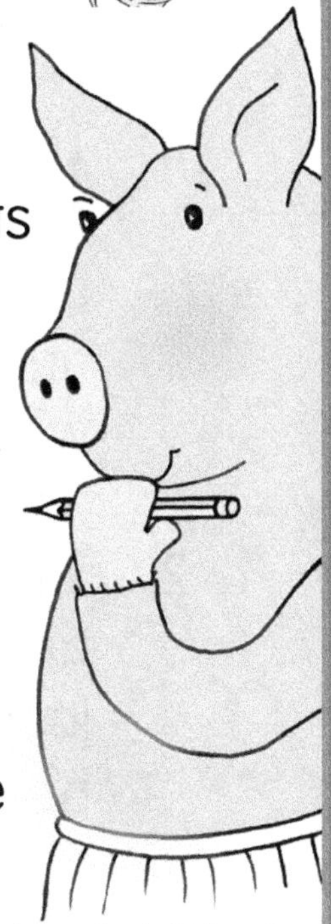

1) Francis picks 360 walnuts and sells five ninths of them to the local bakery. How many walnuts does he sell?

2) There are 18 students enrolled in Jo's art class. One sixth are away on Friday. How many students are present?

3) Mum makes fudge cake and cuts it into 28 pieces. Phoenix eats one seventh of the fudge cake. How many pieces does he eat?

Card 6

Fractions

1) Noah brought 36 biscuits to share with his class. If he has already given out three ninths of them, how many is this?

2) There are 90 tickets for the school play. If two thirds of these seats have been sold, how many tickets are still available?

3) Ethan has 40 rugby magazines. He gives one tenth of them to his little brother. How many rugby magazines does he have left?

Card 7

Fractions

1) There are 42 students in Sue's swimming group. One seventh of the group are already in the pool as they are having the first turn. How many students are still waiting for their turn?

2) Riley picks 180 apples from his orchard and gives four ninths of them to the local school. How many apples does he give away?

No Nonsense Maths Problems Book 2 Year 4

Card 8

Fractions

1) Sarah has to wash 24 glasses. One eighth of them are cracked. How many glasses is this?

2) Tom has saved $240. He spends three eighths of his savings on a computer game and two sixths on a rugby ball. How much does he spend on each item?

3) There are 56 coloured pencils in a packet. Charlotte has to sharpen 28. What fraction of the pencils does she sharpen?

No Nonsense Maths Problems Book 2 Year 4

Card 9

Fractions

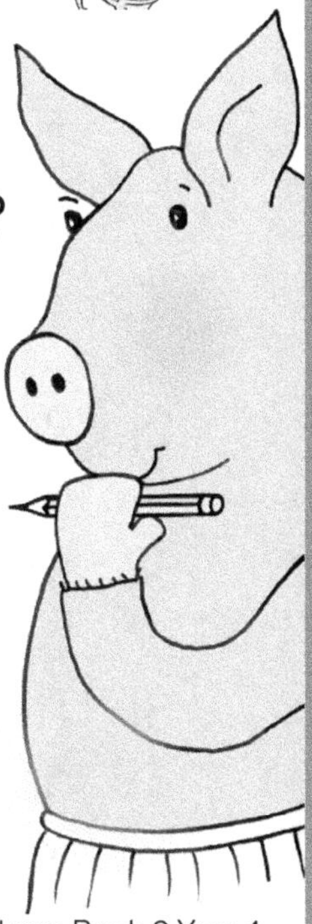

1) There are only 6 boys in Xavier's class of 30 students. What fraction of the class is boys?

2) A farmer has 64 cows and seven eighths of them have had calves this spring. How many cows are yet to calve?

3) There are 18 students in Levi's drama class. One ninth have forgotten to finish their homework. How many students have completed their homework?

Card 10

Fractions

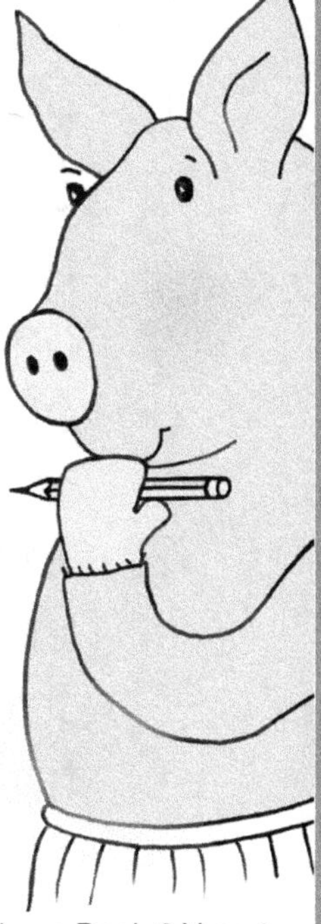

1) Two thirds of the 24 students in Kristine's class have left their homework at home. How many students is this?

2) There are 36 seats on the school bus. 24 seats are taken by boys. What fraction of the total number of seats is this?

3) Dad cooks 35 sausages on the barbecue. He burns one fifth of them. How many sausages does he burn?

No Nonsense Maths Problems Book 2 Year 4

Card 11

Fractions

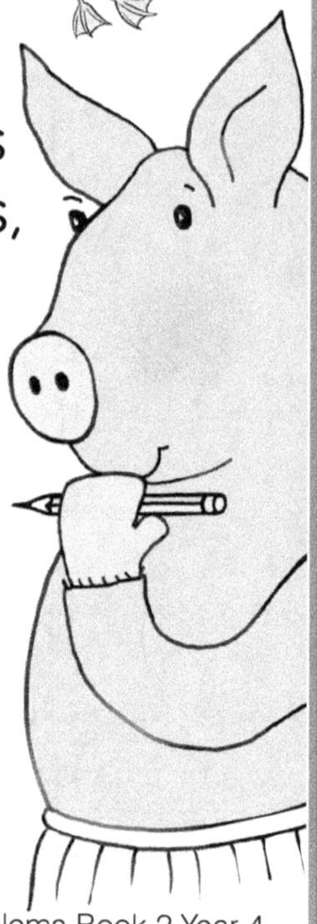

1) There are 48 children queuing up to get tickets for the hydroslide. If 36 of the children are girls, what fraction of the children is boys?

2) Rob picks 180 apples from his orchard and gives 80 of them to the local school. What fraction of the apples does he give away?

3) There are 64 students heading up a mountain on a ski trip. One eighth of them are suffering from travel sickness. How many is this?

No Nonsense Maths Problems Book 2 Year 4

Card 12

Fractions

1) Rob is helping clean up his local beach. He and his neighbours collect 24 bags of rubbish in total. 4 of the bags are filled with only paper. What fraction of the total number of filled bags is this?

2) Jack and Martha both eat some lollies from a bag they have bought to share. Jack eats one third and Martha eats one fifth of the lollies. Who has eaten more?

No Nonsense Maths Problems Book 2 Year 4

Card 13

Fractions

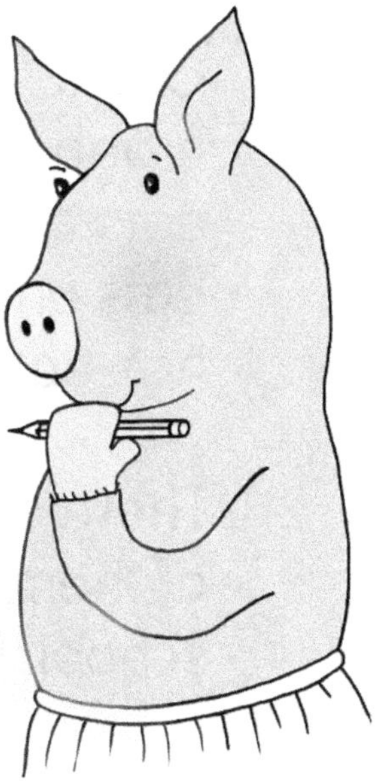

1) Luke has saved $21. If he spends three sevenths of his savings on a computer game, how much does he spend?

2) There are 18 strawberries in a box. If Elliot eats 12 of them, what fraction of strawberries is left?

3) Leo has a 4-litre bottle of olive oil. He pours 1 litre of it into a jug. What fraction of the total amount of olive oil is this?

No Nonsense Maths Problems Book 2 Year 4

Card 14

Fractions

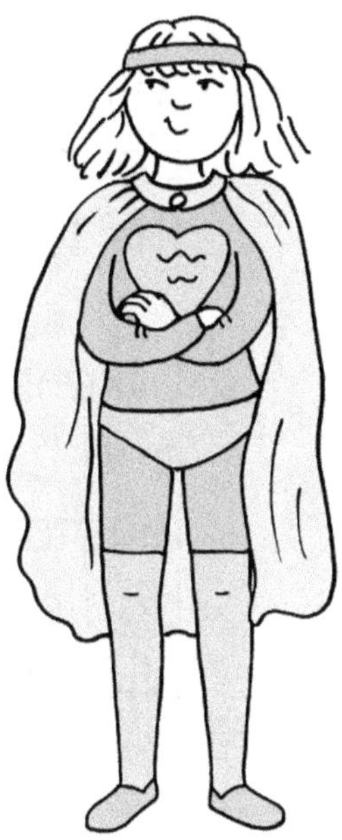

1) There are 36 children in Naomi's class. Four ninths of the children are boys. How many girls are in the class?

2) Thea has 6 oranges. She cuts them into quarters to take them to netball. How many pieces of orange does she take?

3) There are 72 members in Jim's tennis club. One eighth of them have blue racquets. How many members is this?

No Nonsense Maths Problems Book 2 Year 4

Card 15

Fractions

1) Cedric, Jade and John eat some sweets from a bag they have bought to share. Cedric eats one tenth, Jade eats one fifth and John eats one half. Who has eaten the most? Who has eaten the least?

2) There are 36 students in Room 12. One quarter of them forget to bring their homework to school. How many students is this?

No Nonsense Maths Problems Book 2 Year 4

Card 16

Fractions

1) Min Joo has a 5-litre bottle of olive oil. She pours 1 litre of it into a jug. What fraction of the total amount of olive oil is this?

2) Luke is helping clean up his local beach. He and his neighbours collect 24 bags of rubbish in total. 6 of the bags are filled with only glass bottles. What fraction of the total number of filled bags is this?

No Nonsense Maths Problems Book 2 Year 4

Answers 1–12

Addition and subtraction

Card 1
1) 1 586
2) 37

Card 2
1) $2 993
2) 84

Card 3
1) 69
2) 25

Card 4
1) 59
2) $28
3) 98

Card 5
1) $302
2) $261

Card 6
1) $62
2) 391
3) 69

Card 7
1) 167
2) 442
3) 65

Card 8
1) 198 km
2) 55
3) 114 km

Card 9
1) 47
2) $165
3) 27 km

Card 10
1) $1
2) 59

Card 11
1) 150

Card 12
1) 140
2) 245

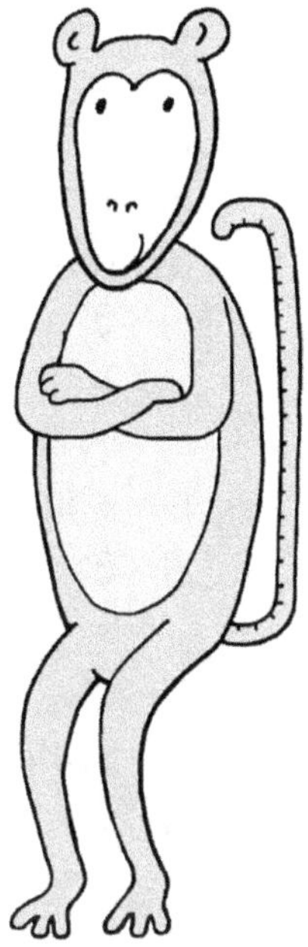

No Nonsense Maths Problems Book 2 Year 4

Answers 13–18

Addition and subtraction

Card 13
1) $783
2) $137
3) $375

Card 14
1) $63
2) 130

Card 15
1) $356
2) 187
3) $78

Card 16
1) $71
2) $556
3) $266

Card 17
1) $308
2) $106
3) $169

Card 18
1) 621 km
2) $35

No Nonsense Maths Problems Book 2 Year 4

Answers 1–11

Multiplication and division

Card 1
1) 4
2) 175 minutes
3) 8

Card 2
1) $24
2) 8

Card 3
1) $144
2) 9

Card 4
1) 35 minutes
2) 8
3) 9

Card 5
1) $700
2) 7

Card 6
1) 25
2) 9 hours

Card 7
1) 13
2) 310
3) 128

Card 8
1) 17, 1
2) $553

Card 9
1) $1 350
2) 22
3) $252

Card 10
1) 29
2) 12
3) 7

Card 11
1) $1 584
2) $3 196

No Nonsense Maths Problems Book 2 Year 4

Answers 12–22

Multiplication and division

Card 12
1) $154
2) $99

Card 13
1) 19
2) 6 km, 378 km

Card 14
1) 4 km, 252 km
2) $770

Card 15
1) 29
2) 7 km, 441 km

Card 16
1) 16
2) $400

Card 17
1) 39
2) $300
3) $164

Card 18
1) 63 km
2) $400
3) 288

Card 19
1) $112
2) 13
3) 248

Card 20
1) $48
2) 4
3) $96

Card 21
1) 305
2) $600
3) 110 km

Card 22
1) 49 km
2) a) 252 km
   b) 1 008 km
3) 30

No Nonsense Maths Problems Book 2 Year 4

Answers
23–34

Multiplication and division

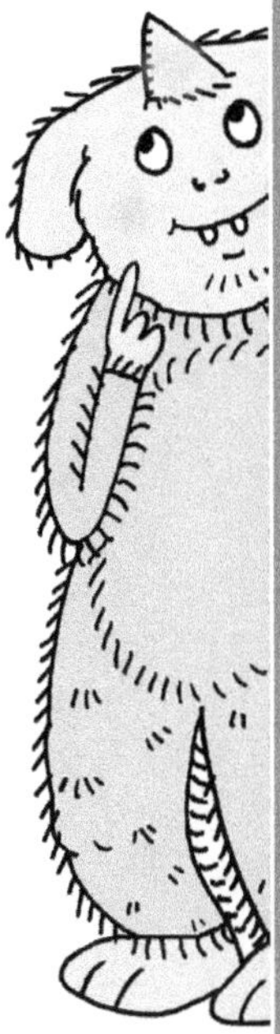

Card 23
1) $500
2) 13
3) $171

Card 24
1) 35 000
2) 12
3) $217

Card 25
1) $256
2) 3 200
3) 8, 3

Card 26
1) 96 000
2) a) 96 kg
   b) 672 kg

Card 27
1) 8 400
2) 4
3) 24

Card 28
1) 1 790
2) $644
3) 7

Card 29
1) 72
2) 6
3) 12, 2

Card 30
1) 108
2) 1 000
3) 56 minutes

Card 31
1) 136
2) 192
3) 10 000

Card 32
1) 7, 2
2) 6
3) 7

Card 33
1) 57
2) 144 kg
3) 410

Card 34
1) 17
2) $126
3) 60

- -

Answers
1–9

Using more than one operation

Card 1
1) 3
2) $18

Card 2
1) 152
2) 120, 6

Card 3
1) 18, $118
2) $1

Card 4
1) $38
2) 34

Card 5
1) $29
2) 266

Card 6
1) $77
2) 27

Card 7
1) $1
2) 629

Card 8
1) 45
2) 18

Card 9
1) 9
2) 147

Answers 10–16

Using more than one operation

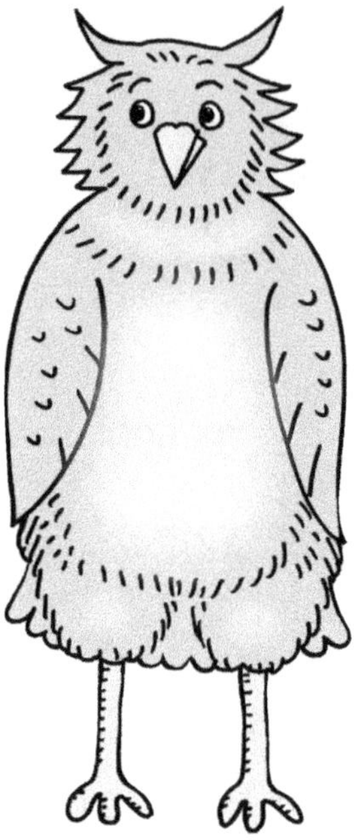

Card 10
1) 14
2) $30

Card 11
1) $37.80
2) 39

Card 12
1) $62
2) $102

Card 13
1) $99
2) $50

Card 14
1) 10
2) $94

Card 15
1) $15
2) $100

Card 16
1) $4
2) 114

No Nonsense Maths Problems Book 2 Year 4

Answers 1–12

Fractions

Card 1
1) a) 42
   b) 14
2) Tim

Card 2
1) $\frac{1}{3}$
2) a) 120
   b) 160

Card 3
1) 600 m
2) $\frac{1}{9}$
3) $\frac{1}{8}$ or 6

Card 4
1) 6
2) 9
3) $90

Card 5
1) 200
2) 15
3) 4

Card 6
1) 12
2) 30
3) 36

Card 7
1) 36
2) 80

Card 8
1) 3
2) $90, $80
3) $\frac{1}{2}$

Card 9
1) $\frac{1}{5}$
2) 8
3) 16

Card 10
1) 16
2) $\frac{2}{3}$
3) 7

Card 11
1) $\frac{1}{4}$
2) $\frac{4}{9}$
3) 8

Card 12
1) $\frac{1}{6}$
2) Jack

No Nonsense Maths Problems Book 2 Year 4

Answers 13–16

Fractions

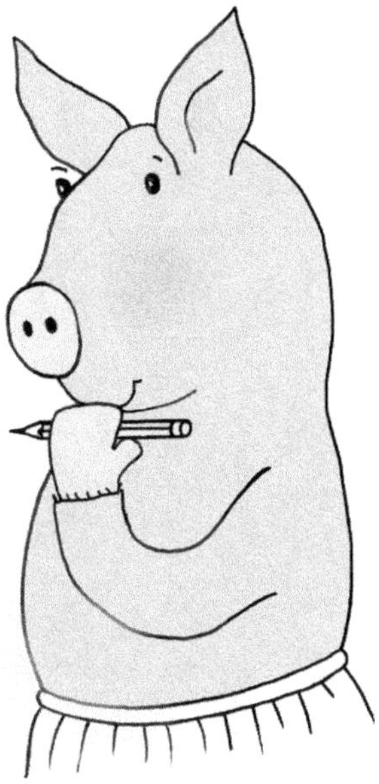

Card 13
1) \$9
2) $\frac{1}{3}$
3) $\frac{1}{4}$

Card 14
1) 20
2) 24
3) 9

Card 15
1) John, Cedric
2) 9

Card 16
1) $\frac{1}{5}$
2) $\frac{1}{4}$

No Nonsense Maths Problems Book 2 Year 4

www.ingramcontent.com/pod-product-compliance
Lightning Source LLC
LaVergne TN
LVHW062319051126
851857LV00015B/1575
* 9 7 8 1 9 2 7 1 9 0 4 3 2 *